CW01640420

PAGE

Following the death of Bob Ramsay in the summer of 2000, the information has been revised and updated by *Waverley*'s Chief Engineer, Ken Henderson. Details are now included of the Cochran Thermax boilers fitted to the ship during her major rebuild in 2000.

The Triple Expansion Engine was the logical conclusion to previous designs. The simple diagonal, single cylinder engine was followed by the successful compound engine. Indeed, these designs were in operation up to the end of World War II, and after as in the case of ***Lucy Ashton***, ***Maid of the Loch*** and ***Kingswear Castle***.

The principal gain by having three cylinders is that the pressure is lowered a certain amount in each cylinder, and the temperature drop does not take place all at once, but is divided into three stages.

Consider a single cylinder taking steam at 200 pounds per square inch (PSI). The great temperature difference between admission and exhaust would cause excessive condensation owing to the cylinder cooling during exhaust. By dividing this temperature drop between three cylinders, condensation losses are proportionally decreased, and are counterbalanced by re-evaporation. It should be considered that a drop in temperature corresponds to work lost.

If the steam pressure at the regulating valve is, say, 180 PSI, the High Pressure (HP) engine receiver would be 170 PSI, the Medium Pressure (MP) receiver 58/60 PSI and the Low Pressure (LP) one PSI, these being gauge pressures. The final exhaust from the LP to the condenser is at 25 inches vacuum, or $2\frac{1}{2}$ PSI absolute. It will be noted that some pressures are given in gauge and some in absolute terms. A comparison is shown here. Normally gauges measure pressure relative to atmospheric pressure at approximately 15 PSI Absolute.

Gauge Pressure (PSI)	15	10	5	0	–2·5	–5	–7·5	–10	–12·5	–15
Absolute Pressure (PSI)	30	25	230	15	12·5	10	7·5	5	2·5	0
Vacuum (Inches Mercury)	–	–	–	0	5	10	15	20	25	30

The cylinder arrangements are governed by the space available, a typical design being the HP in the middle, with the MP and LP at either side of the engine room. The disadvantage is the necessity of interconnecting pipes between the cylinders, especially the large pipe from the MP, across the HP, to the LP unit. In mitigation, the general arrangement permits wider deck space on either side of the engine room. Placing the two larger cylinders either side of the smaller will help with athwart ship weight balance.

The alternative arrangement is to have HP, MP and LP in series. This would be the traditional layout for an "Up and Downer" placed fore and aft driving a propeller. The cross-over pipes are eliminated, and steam flows to and from each cylinder by passages moulded within the castings.

P.S. *WAVERLEY* – DIAGRAMMATIC ARRANGEMENT OF MAIN ENGINE OPERATING GEAR (running position)

by Rankin & Blackmore, Greenock

Triple Diagonal:
Cylinders 24", 39", 62"; 66" stroke
Stephenson Valve Motion

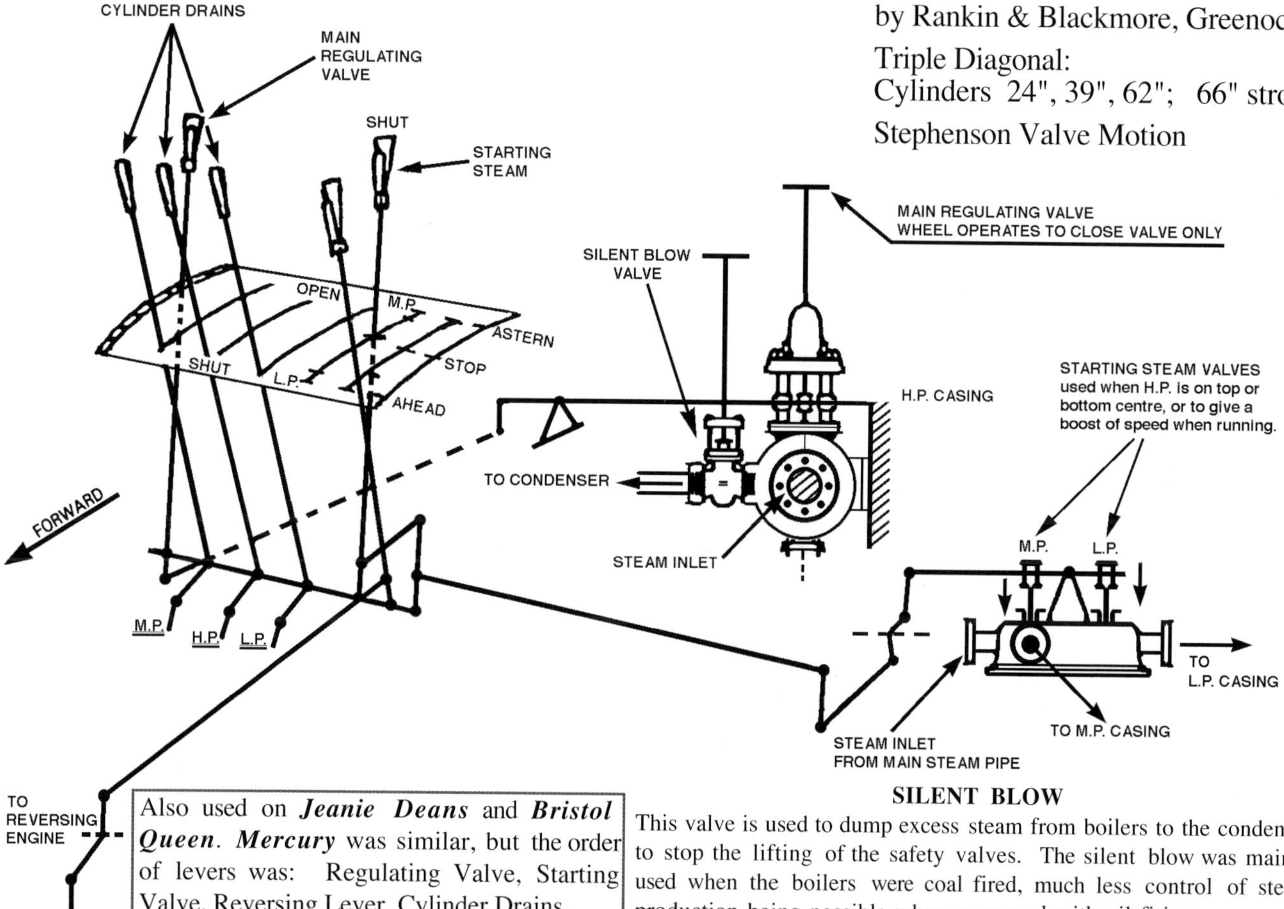

Also used on ***Jeanie Deans*** and ***Bristol Queen***. ***Mercury*** was similar, but the order of levers was: Regulating Valve, Starting Valve, Reversing Lever, Cylinder Drains.

SILENT BLOW

This valve is used to dump excess steam from boilers to the condenser to stop the lifting of the safety valves. The silent blow was mainly used when the boilers were coal fired, much less control of steam production being possible when compared with oil firing.

This design, while decreasing the pressure and temperature drop between the units, is more complicated to manufacture, and thus more expensive. The ***Jupiter*** and ***Juno*** of 1937 had this arrangement.

Pressure relief valves are fitted at each end of the cylinders and on the MP and LP valve chests. Relief valves are used to limit pressures to below the designed maximum pressure. Drain cocks or valves operated from the control platform are fitted to remove the water of condensation from each end of the cylinders. These drains are led to the condenser. Additional drains are fitted to the valve chests. These are normally only used when warming through the engine in the morning, and will be led to the bilge.

The engine speed is controlled by the Regulating Valve, operated by a lever which opens or closes the valve as required. The reversing or direction lever operates the valve gear for ahead or astern running via a reversing engine which is required to move the heavy gear to the desired position. In emergency, the gear can be "jacked over" by means of a hand pump operating on the hydraulic cylinder of the reversing engine.

If the HP crank is on top or bottom centre, the engine will not turn when steam is admitted. To overcome this situation, impulse or starting valves are fitted. These valves, often operated by a single lever, receive direct steam from the boilers; one valve leads to the MP valve chest, the other to the LP valve chest. The MP or LP valve will then control the steam cut off to the cylinder as normal. Thus with the HP on a centre, the other cranks will be off centre, and by admitting direct steam to either the MP or LP cylinder, the shaft will turn. The impulse is also used when warming through, or to boost the speed when running by passing direct steam, usually to the MP cylinder.

The port side gauge board consists of the Main Steam Gauge (graduated zero to 350 PSI), HP Receiver (zero to 350), MP Receiver (zero to 150), LP Receiver (compound gauge, zero to 50 PSI and zero to 30 inches vacuum) and Condenser Gauge (zero to 30 inches vacuum). On the starboard side gauge board are the Boiler Pressures (zero to 250 PSI) and Auxiliary Exhaust (compound, zero to 50 PSI and zero to 30 inches vacuum).

P.S. *CALEDONIA* (1934) – DIAGRAMMATIC SKETCH OF OPERATING STAND

by Wm. Denny, Dumbarton

Triple Diagonal:
Cylinders 20", 31·5", 50"; 60" stroke

Brock Valve Motion

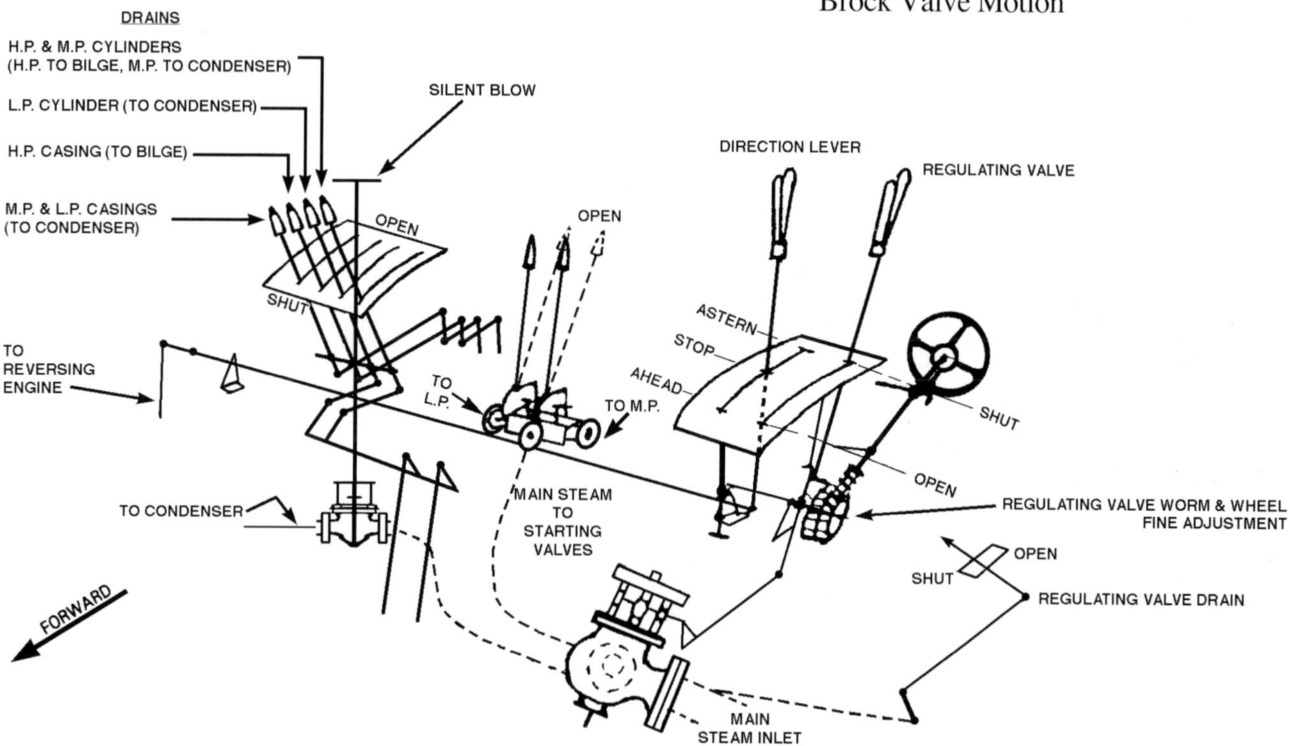

PADDLE ENGINE WITH STEPHENSON VALVE MOTION
(e.g. *WAVERLEY*)

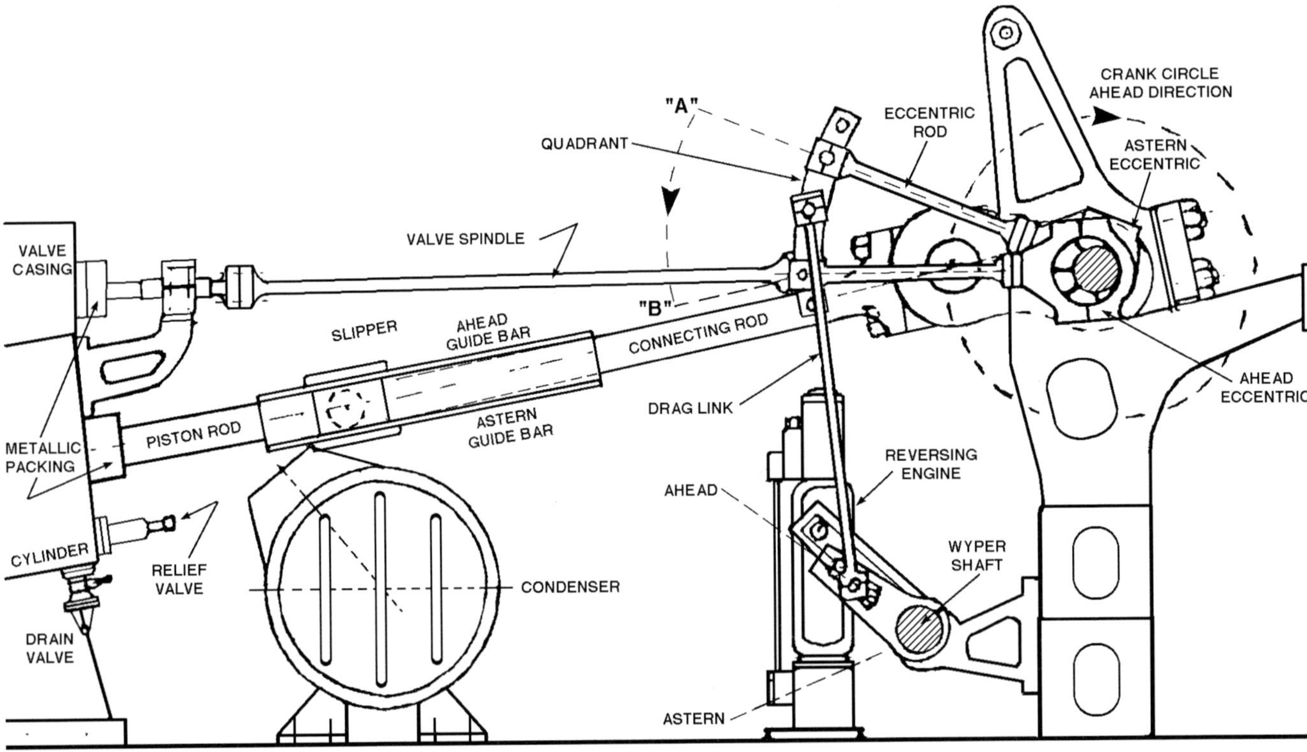

Valve Gear shown in the "Ahead" position. To go astern, the Reversing Engine, operated from the Control Station, moves the Quadrant down through arc AB. The "Astern" Eccentric Rod thus comes into line with the Valve Spindle, so changing the valve position to the "Astern" setting.

BROWN'S COMBINED STEAM AND HYDRAULIC REVERSING ENGINE

The engine consists of a steam cylinder and slide valve, with a hydraulic cylinder whose piston is secured to the common piston rod. The engine is connected to the lever "A" secured to the Reversing (Wyper) Shaft. The motion from the piston rod is transmitted by a slipper working in a slot as shown.

If the Reversing Control Lever is moved "ahead", the valve spindle is moved upwards. Steam enters the cylinder bottom, moving the steam and hydraulic pistons upwards. Water is displaced from the top of the hydraulic cylinder via pipe "B", and through the hydraulic valves to the opposite end of the cylinder. This gives a slow, smooth motion. At the end of the stroke, the linkage shuts steam off by returning the slide valve to the closed position. The Reversing Shaft moves the Valve Gear into position for operation by the Ahead Eccentric. Water in the hydraulic tank will make up any losses in the system.

The opposite sequence, i.e., pistons to the bottom of the cylinders, would run the main engine astern.

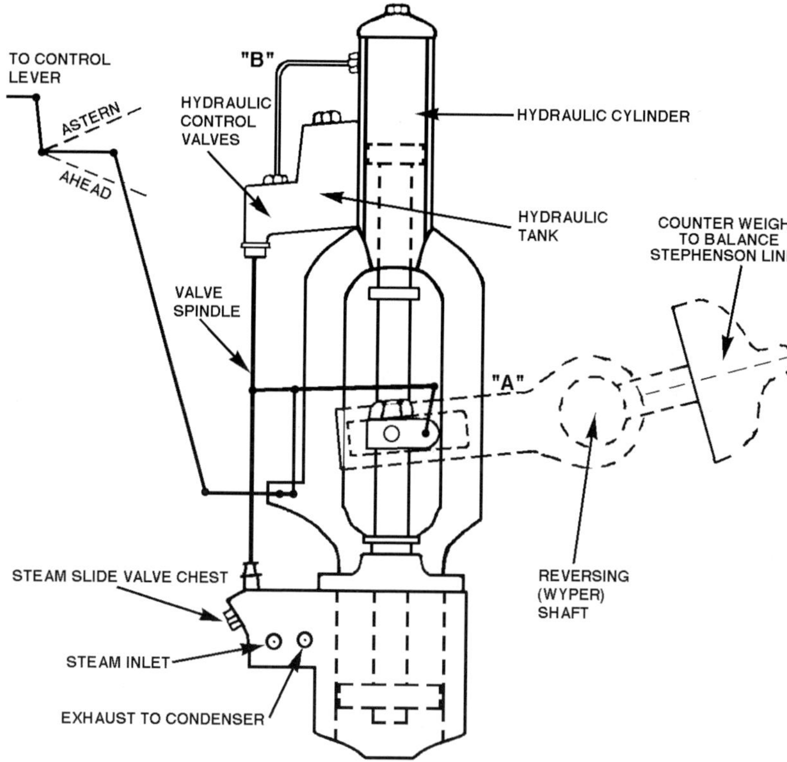

TRIPLE EXPANSION PADDLE MACHINERY
CYLINDER VALVE TYPES AND ARRANGEMENTS

The following applies to equipment fitted to steamers including ***Jeanie Deans***, ***Mercury***, ***Bristol Queen*** and ***Waverley***.

HP CYLINDER – PISTON VALVE (INSIDE STEAM ADMISSION)
Steam Pressure 180 PSI

The piston is commencing its outward stroke, and the Piston Valve is inwards, just commencing to open steam to the cylinder behind the piston. Steam is exhausting from the front of the piston, over the ends of the piston valve, via pipes to the MP Valve Chest.

Exhaust Pressure:- 60 PSI (approximately)

Advantages:
1. Light weight valve.
2. Steam at inside of valve gives a balancing effect, taking weight off the valve motion gear.
3. Exhaust steam outside of valve takes pressure away from spindle gland, which is thus subject to 60 PSI exhaust pressure only.
4. Reduced friction.
5. Reduced travel.

Disadvantages:
1. Two exhaust pipes necessary to conduct the steam to the MP Cylinder.
2. Liable to leakage past rings.

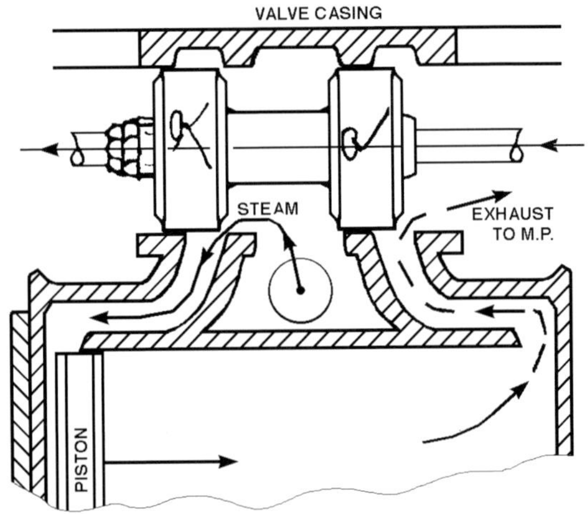

HP PISTON VALVE
(Inside Steam)
Piston leaving top centre.
Valve just opening to top steam,
and exhausting bottom steam.
Ahead eccentric follows the crank.

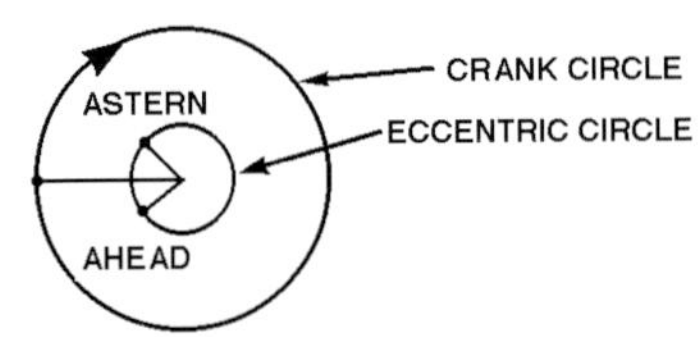

MP CYLINDER – PISTON VALVE (OUTSIDE STEAM ADMISSION)

This valve is similar to that on the HP Cylinder, except that steam is on the outside of the valve and exhaust on the inside. Only one exhaust pipe is thus required to lead the steam to the LP Cylinder at approximately 1 PSI. Owing to the large dimensions of this pipe, fitting two as in the HP case would be impractical.

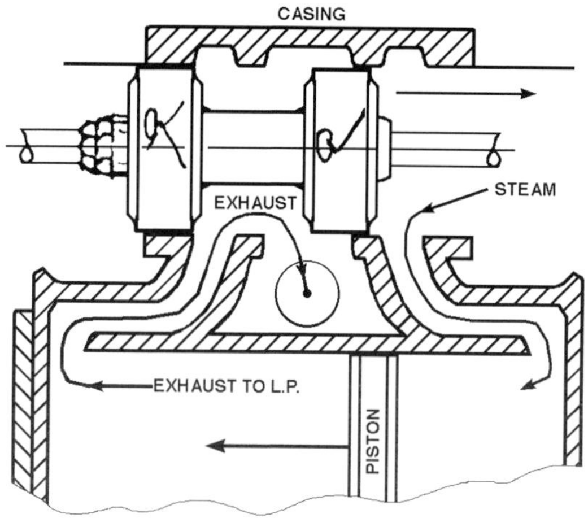

MP PISTON VALVE
(Outside Steam)
Ahead eccentric leads the crank.

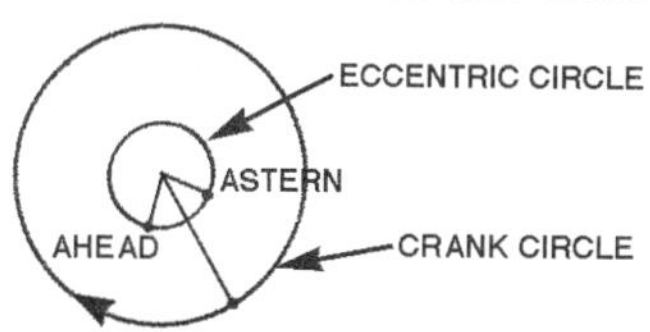

LP CYLINDER – SLIDE VALVE

A large diameter Piston Valve has the disadvantage of increasing the length or width, in the case of paddle engines. Thus a Balanced Slide Valve is usually fitted to the LP Cylinder.

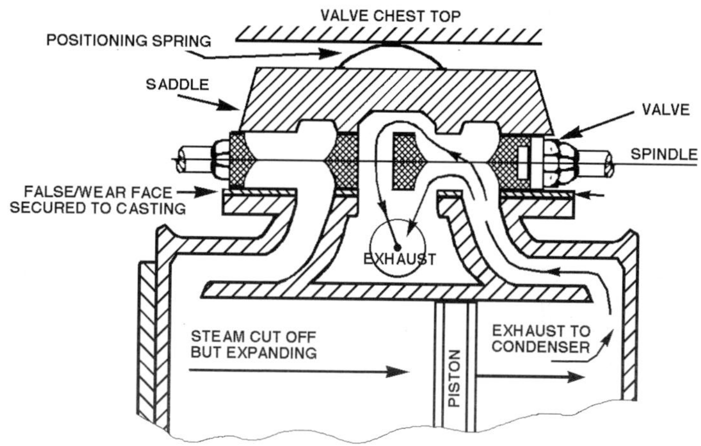

LP VALVE (Andrews & Cameron)
Ahead eccentric follows the crank.

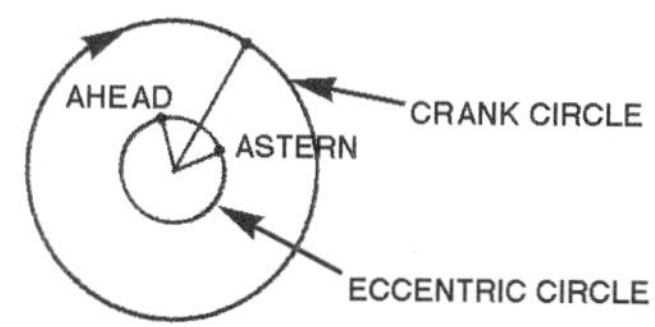

BROCK'S VALVE GEAR
(Wm. DENNY, DUMBARTON)
e.g., ***Caledonia*** (1934)

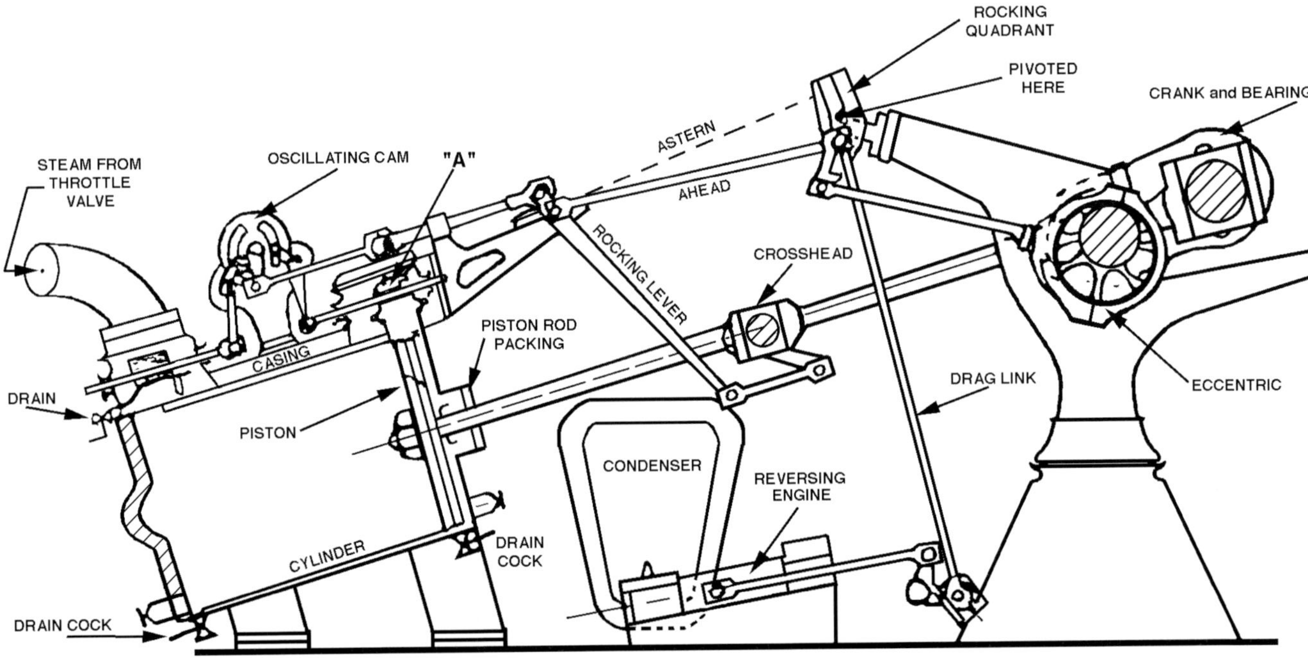

When P.S. CALEDONIA was destroyed by fire in 1980, her engine was salvaged, and is now preserved in operational condition in the steam museum at Hollycombe House, Liphook, Hampshire.

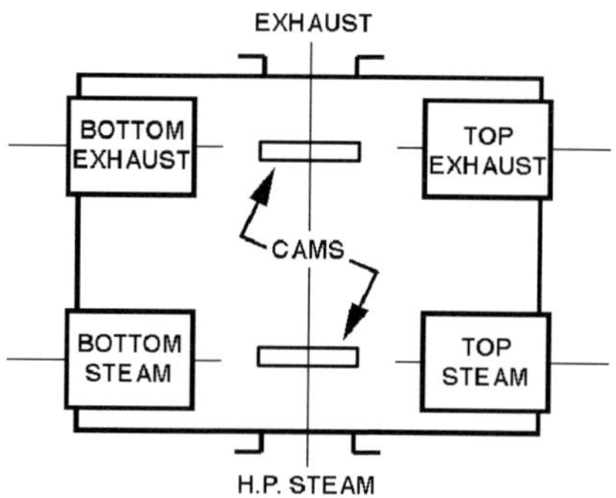

ARRANGEMENT OF VALVES ABOVE EACH CYLINDER

(Each valve can be set separately.)

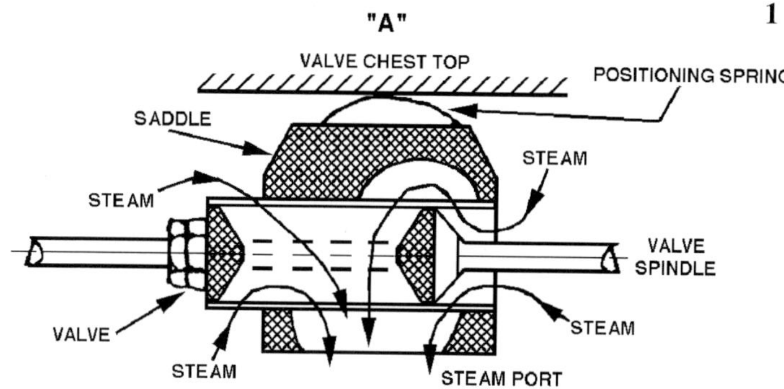

ANDREWS & CAMERON VALVE

Small light-weight valve, with four openings, thus short travel, and less friction. The exhaust valve is similar, but with the flow reversed.

MATERIALS

Cylinders:	Close grained cast iron
Piston Rods:	Forged steel
Crossheads:	Forged steel, with bronze-white metal lined shoes
Connecting Rods:	Forged steel, with double bronze bearing bushes at fork end and single bronze-white metal lined bearing at crank end
Shafting:	Hydraulic forged, of Siemens mild steel, to D.O.T. requirements. The shaft is hollow, of the built-up type. Cranks will be set at 90° to each other for a 2-cylinder engine, and 120° for a 3-cylinder engine. It should be noted that a hollow shaft is stronger than a corresponding solid one. In the latter, impurities tend towards the centre, and defects can spread outwards. The hollow shaft has all these impurities removed.
Frame (Entablature):	Cast steel, with forged steel guides and columns.

EXPANSION GEAR

The purpose of this gear is to alter the piston or slide valve travel in order to adjust the power and revolutions of the engine. As shown, it is incorporated in the reversing shaft crank, one for each cylinder. It consists of a screw and a nut or block on which the drag link is free to swing. The nut is free to move within the slot from one end to the other, by rotation of the handle.

It will be seen by the position of the nut in sketch "A" that the valve spindle is in line with the eccentric rod, the travel being equal to the eccentric throw. The valve travel is normal, giving the designed cut-off at 75%. Sketch "B" shows the block screwed down towards the Wyper shaft. This action lowers the quadrant so that the valve spindle, being no longer in line with the eccentric, operates at reduced travel, the effects being as shown. Sketch "C" has the block screwed out past the in-line position, raising the quadrant, and leaving the valve spindle overhung. This position gives valve travel greater than the eccentric throw, the effects being as shown.

While both the MP and the LP cylinders have similar gear, the adjustment of the HP expansion affects the whole engine's power and fuel consumption. Any alteration to the MP or LP is to increase or decrease the power contributed by that cylinder to give a satisfactory balance. It should be noted that in the astern position shown, no matter what the expansion setting may be, the valve spindle is always in line with the eccentric to give normal full power in that direction.

In practice, where there is constant manœuvring, the HP expansion is kept at "link-in-line" for quick response to telegraph movements. During "Half Ahead" cruises, it is economical to "link in" all cylinders by means of the reversing gear. In long coastal passages it is an asset to link in using the HP expansion.

The main reason for linking in an engine is to reduce speed by making the steam cut-off earlier in the stroke, thus using the steam more expansively in that cylinder. Reducing speed by shutting in the regulating valve throttles steam to the HP chest, resulting in a pressure drop and a loss in the economy of high pressure steam.

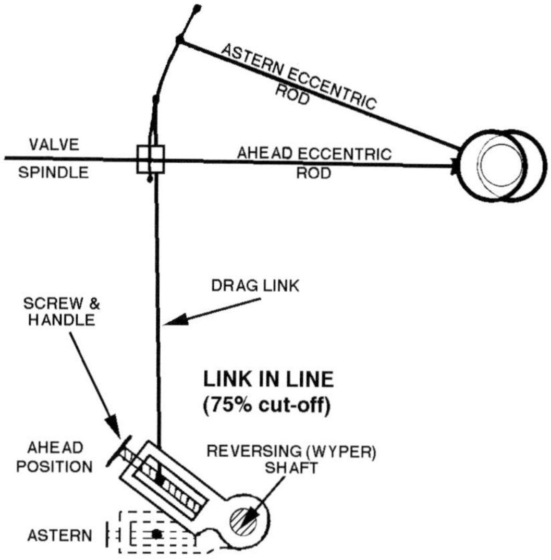

"A" – Link in line

Valve spindle is in line with eccentric rod, thus valve stroke or travel is ***equal*** to the eccentric throw.

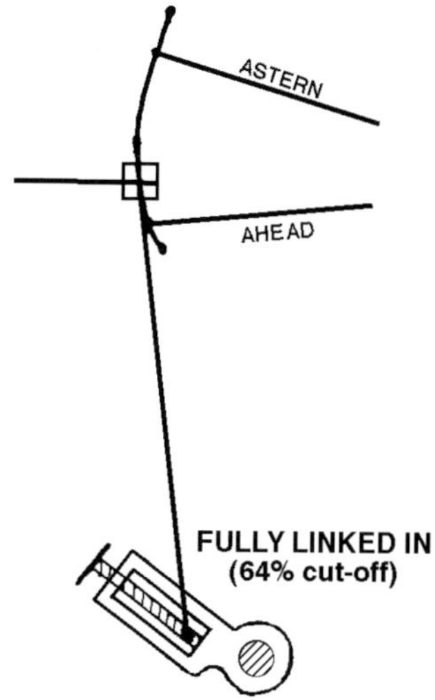

"B" – Gear Linked In

Valve travel is ***less*** than the eccentric throw.

Effects: Linked in HP
Steam is shut off earlier, valve travel is decreased, giving –

1. reduced total power
2. decreased revolutions
3. fuel consumption decreased.

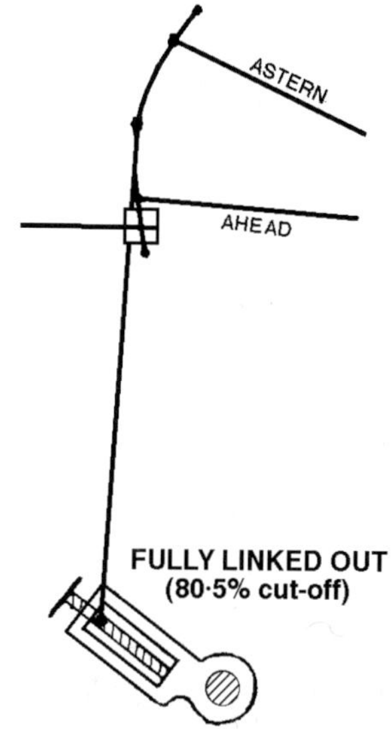

"C" – Gear Linked Out

Valve travel is ***greater*** than the eccentric throw.

Effects: Linked out HP
Steam is cut-off later, as valve travel is increased, giving –

1. total power increased
2. revolutions increased
3. fuel consumption increased.

BOILERS

1. SCOTCH BOILER

This type was used on high powered steamers, paddle and turbine, being of double-ended design, as illustrated. It was a robust boiler with a large steaming capacity. Some details now follow.

Shell: Of mild steel plating, frequently in three rings in two halves, joined by two treble riveted double butt-strap joints. The tensile strength of the plating is 28-32 tons per square inch.

End Plates: May be built up with several plates, flanged inside the shell, with double riveted circumferential seams.

Combustion Chambers: Some combustion takes place in the furnaces, and the remainder occurs in the combustion chambers. The products of combustion return to the uptake via plain and stay tubes.

Furnaces: Always made of corrugated steel with welded seams. The corrugations give, for a given thickness, additional strength with horizontal flexibility; at the same time the heating surface is increased compared to that of a plain furnace. The number of furnaces is dependent upon the boiler diameter. Furnaces are flanged at the combustion chamber end and shaped for easy withdrawal through the front plate. This arrangement assists renewal if required.

Tubes: The front tube plate and combustion chamber plate are tied together by stay tubes. These are screwed through both tube plates, the material being wrought iron or steel. Plain tubes form the major part of the heating surface and can be of seamless steel, expanded into each tube plate.

Stays: Longitudinal stays between the end plates are secured by nuts on either side of these plates. Screwed stays are used for staying chamber tops, sides and back. The material is steel for longitudinal stays, and iron or steel for the others.

DOUBLE ENDED SCOTCH BOILER

Length 18'6", Mean diameter 14'0"
200 PSI, 8 Fires

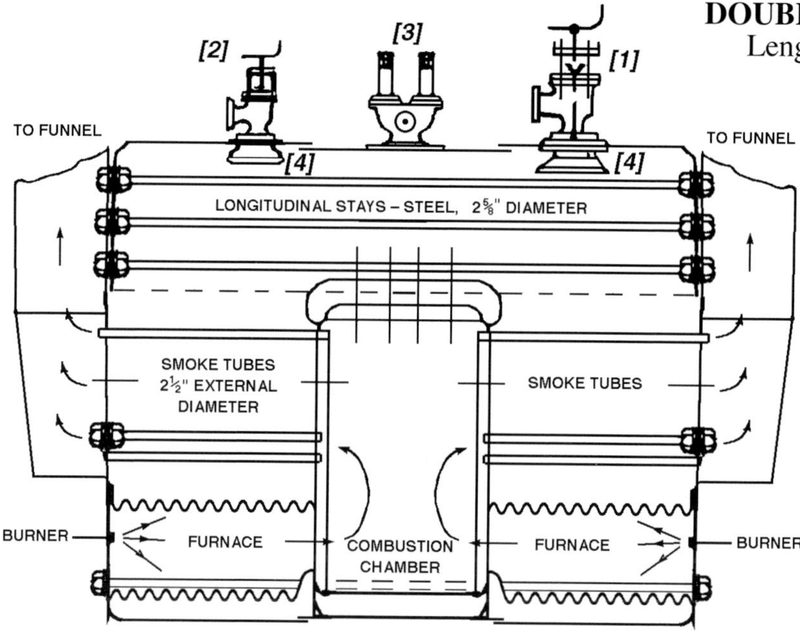

SIDE SECTION, SHOWING PATH OF FLAME & GASES

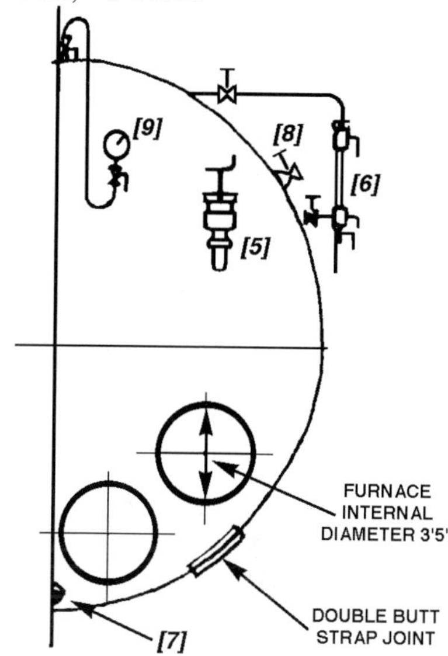

HALF END VIEW

1. Main Steam Stop Valve
2. Auxiliary Steam Stop Valve
3. Safety Valves
4. Steam Separators, Main & Auxiliary
5. Main Feed Water Check Valve
   Auxiliary Feed Water Check Valve on other side
6. One Water Gauge Glass & Mountings;
   Second Water Gauge Glass & Mountings on other end
7. Bottom Blow Down Valve } To overboard
8. Scum Valve } To overboard
9. Pressure Gauge

NOTE: Feed checks are shown in diagrammatic positions only, and would in practice be located at the boiler side.

2. NAVY BOILER

This design was used in gunboats, and resembles the locomotive boiler. It served the Clyde paddlers from the end of last century right up until the 1934 ***Caledonia***, and indeed, when she was re-boilered in 1955, the same type was installed. One advantage is that their size allows two to be fitted if desired, thus giving a "back-up" should one fail. The steamer would then operate at reduced speed on one.

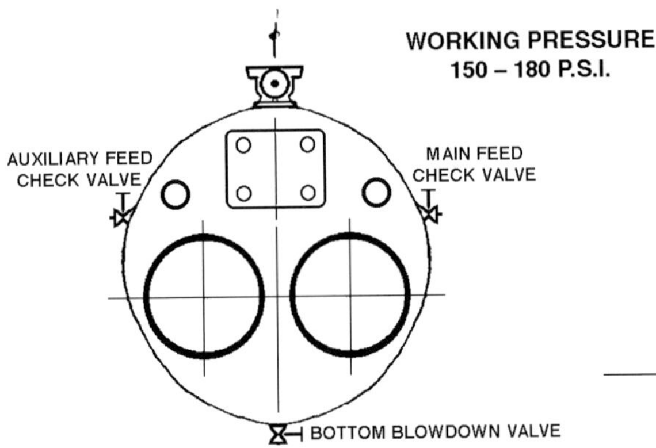

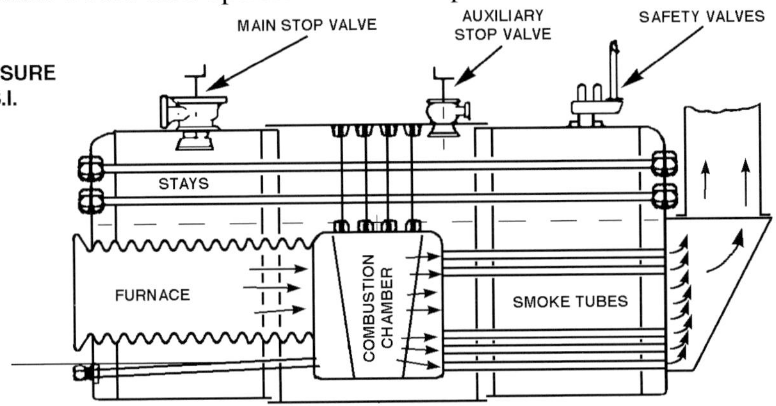

MATERIALS
(Scotch, Navy and Steam Bloc Boilers)

Pressure parts:	Siemens-Martin steel
Tubes:	Wrought iron
Factor of Safety:	Not less than 4

GENERAL

Corrugated Furnaces:

1. Stronger than a plain furnace of the same dimensions
2. Better expansion allowance by means of the corrugations
3. More surface for the same length, giving better evaporation

Forced Draught: Navy and Scotch Boilers operated under a closed stokehold system. Air was delivered to the stokehold by one or two large diameter steam-driven fans. This air under pressure gave increased draught to the fires, hence better combustion. As the stokehold was pressurised, access by airlocks was necessary, as a decrease in air pressure could cause a dangerous "blow back" from the fires under steam conditions.

3. STEAMBLOC (BABCOCK) BOILER

This is a combined fire-tube and water-tube boiler, the latter tubes being for rapid steam raising.

As is modern practice, welding is used in place of rivets. The furnaces are of the plain type, built in sections joined by Bowling Hoops.

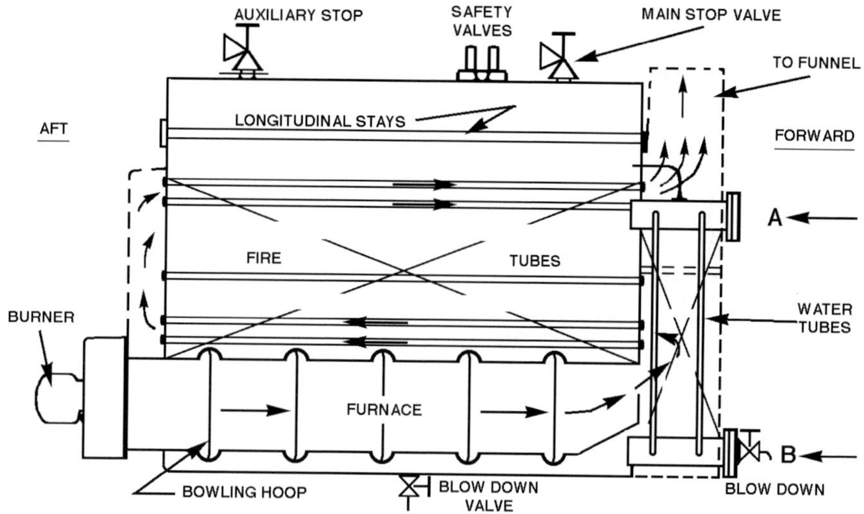

DIAGRAMMATIC ARRANGEMENT STEAMBLOC TWO-FURNACE BOILER

P.S. ***Waverley***
(1981 – 2000)

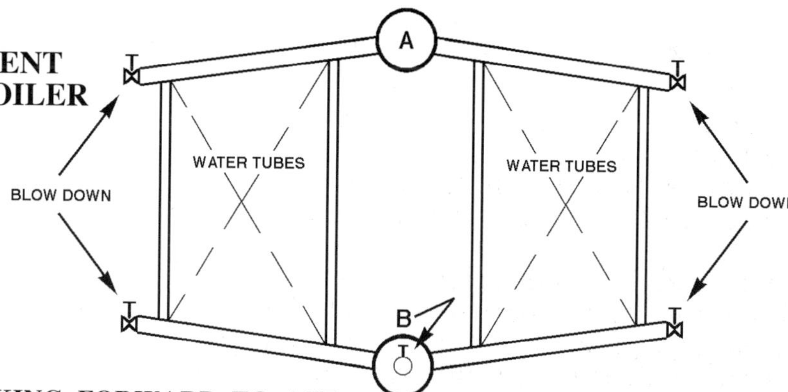

VIEW LOOKING FORWARD TO AFT

4. COCHRAN THERMAX BOILER

(as fitted to *Waverley* in 2000)

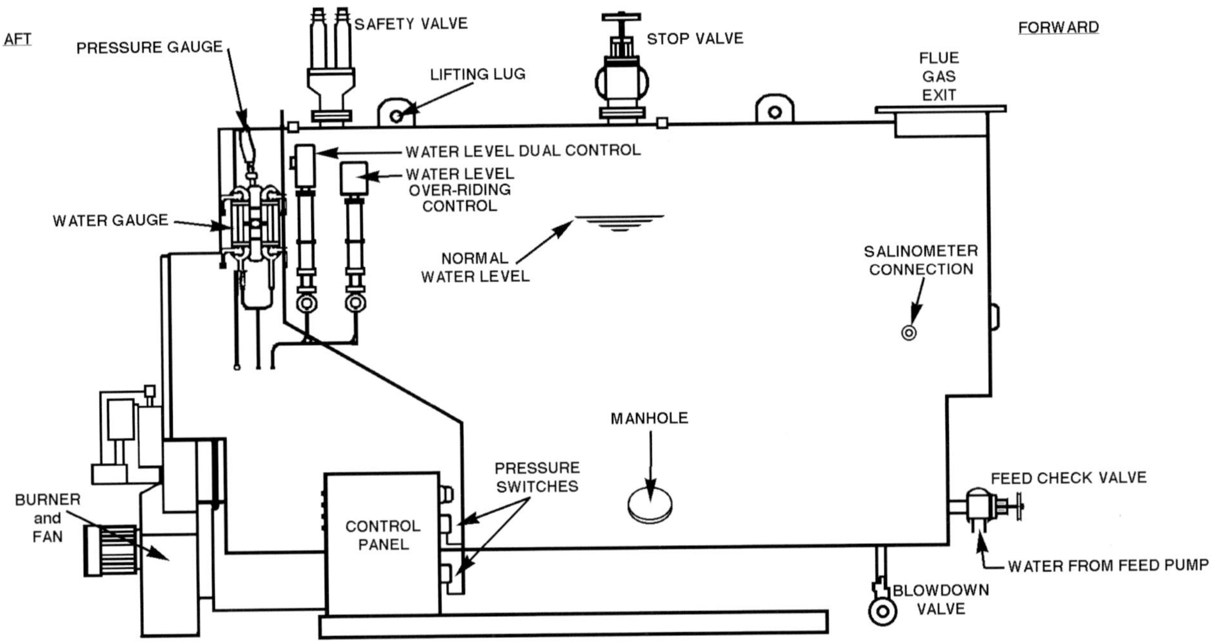

BOILER ARRANGEMENT

In the Millennium Rebuild of 2000, ***Waverley*** was fitted with two Cochran Thermax boilers, whereas she had previously always had one boiler – a double-ended Scotch Boiler from 1947 till 1981, and a Babcock Steambloc Boiler from 1981 till 2000.

The boilers are supplied with water by a feed pump. Water enters the boiler through one of the feed check valves, and when the water reaches normal water level the feed control valve is closed.

SECTIONAL VIEW OF THERMAX BOILER, SHOWING GAS FLOW

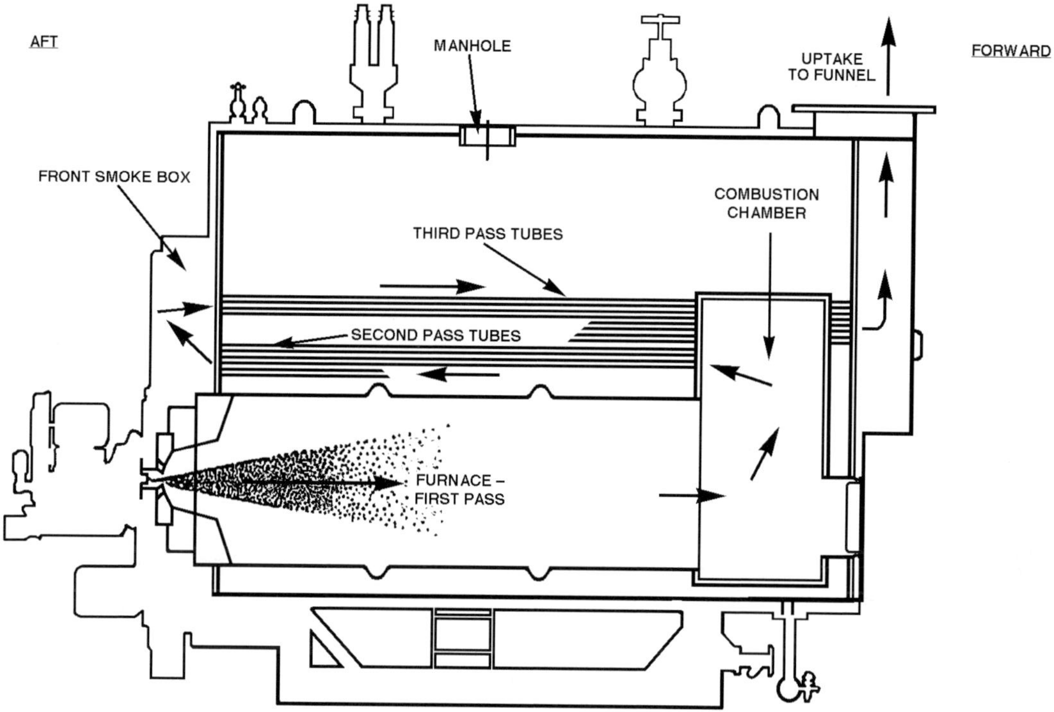

Each boiler is fitted with a rotary cup burner which is bolted to the front casing. Air required for combustion is provided by the forced draught fan, which is integral to the burner unit. The furnace receives radiant heat from the flame and the hot gases from the flame give up heat to the tubes in the second pass as they travel to the front chamber.

Gases turn in the front chamber and give up more heat to the tubes in the third pass as they travel to the rear chamber. The gases leave the boiler by the funnel uptake at the forward end of the boiler.

BOILER FEED WATER SYSTEM

BOILER FEED PUMPS

These pumps, two in number – one working, one standby – draw water from the hotwell and deliver via a filter and heater to the boiler(s).

They are double acting vertical units, with the steam cylinder above as usual. The pump operating at, say, 180 PSI steam pressure, must deliver against a greater pressure, and this is achieved by having the steam piston larger than the pump bucket, e.g., piston 10" diameter, bucket 7". The output from a direct acting pump will conform to the equation $F = P \times A$ where F = Force, P = Pressure and A = Area of piston. As both steam and water pistons are on a common shaft, both forces are equal. Therefore

$$P_{steam} \times A_{steam} = P_{water} \times A_{water}$$

The output/speed of the pump is controlled by a float in the hotwell. The float is connected by levers and linkage to a steam valve which supplies the pump.

If the tank water level rises, the float rises and operates the steam valve to the pump, more steam is given to the pump, which moves faster, bringing the water level back to normal working level. Conversely, if the water level falls, the pump will slow down until the water level becomes normal.

FEED WATER HEATER

This is a vessel similar in construction to the condenser. Exhausts from various pumps and auxiliary engines (except the steering engine and capstan) are collected and passed to the heater. These exhausts are condensed by feed water passing through tubes on the way to the boiler(s) from the feed pump. The steam on the outside of the tubes gives up heat to the feed, is condensed and drains to the hotwell.

The feed may enter the heater at 60/65°C (140/150°F) and leave at 120°C (248°F). In this way hot water enters the boiler, gives easier steaming, with a corresponding reduction in fuel consumption, and reduces internal stresses in the boiler. To obtain these conditions, the exhaust line pressure is regulated to give about 25 PSI, surplus steam being passed to the condenser.

It will be seen that the same feed water is used over and over again, but owing to unavoidable leakages and evaporation losses take place. To make up these, water is stored in separate tanks and introduced to the system via the condenser, or the hotwell tank.

BASIC FEED WATER SYSTEM - DIAGRAMMATIC SKETCH

Temperatures are approximate

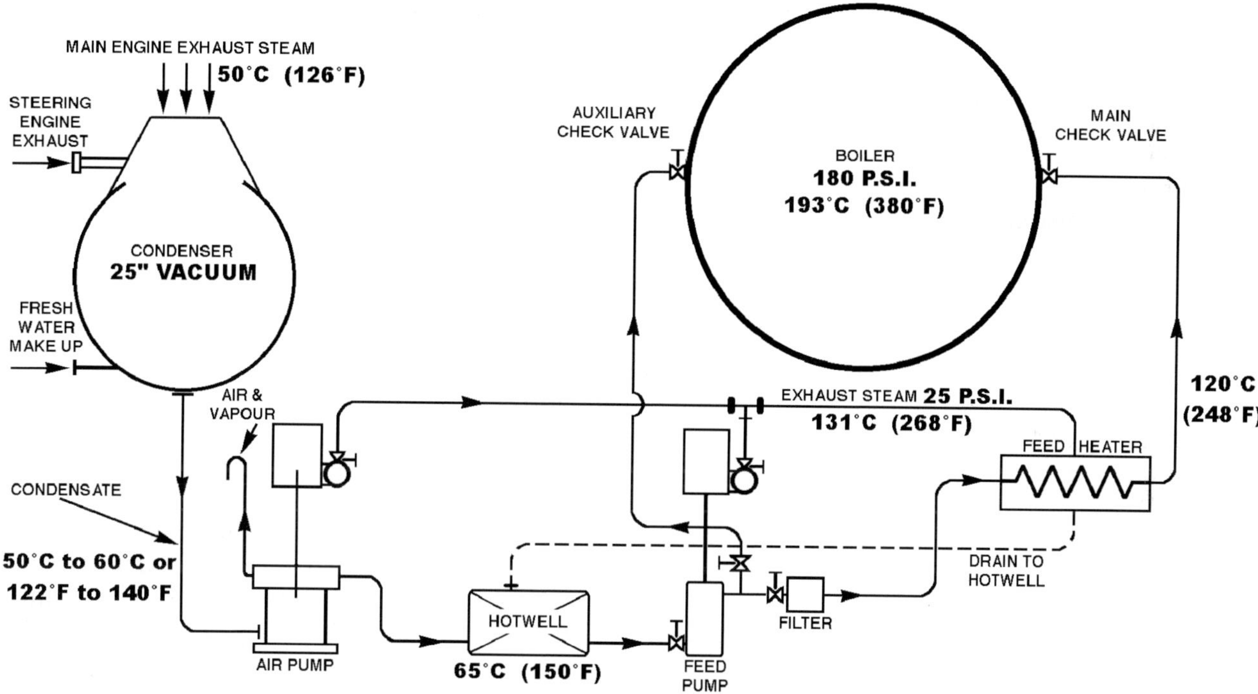

CONDENSER

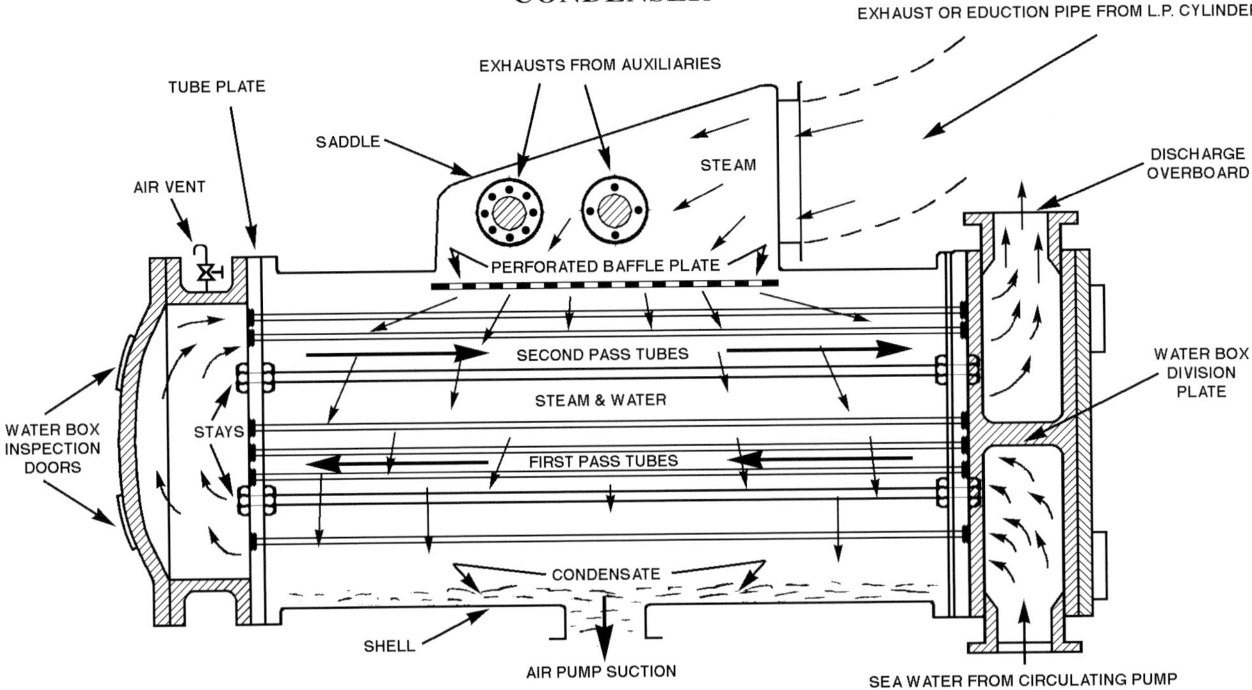

DIAGRAMMATIC SKETCH OF CONDENSER

MATERIALS

Shell:	Steel plates
Saddle:	Steel Plates
Water Boxes:	Cast Iron
Tube Plates:	70% Copper, 35% Nickel
Tubes:	"Alumbro" Metal (76% Copper, 22% Zinc, 2% Aluminium)

CONDENSER

From the diagram, it will be seen that exhaust steam from the main and auxiliary machinery passes over many tubes through which sea water flows from the rotary circulating pump. The steam is thus changed to water (condensate) and falls to the bottom of the condenser, from where it is removed by the air pump.

The condenser tubes are usually #/4" diameter and pass through tube plates to the water boxes. They are made tight against entry of sea water between plate and tubes by special packing.

The cooling surface averages 0·7 square feet per horse power, approximately, so ***Waverley*** being 2100 HP requires 1470 square feet. In practice it is 1600 square feet allowing for exhaust from auxiliary engines.

The condenser pressure gauge indicates the vacuum in inches of mercury (Hg). If the gauge indicates 25" with a barometer reading of 30", then the condenser pressure is 5", or 2½ PSI absolute. This is the back pressure in the condenser into which the engine exhausts. If there was no vacuum, only atmospheric pressure at 15 PSI, the engine would slow down when operating against that pressure.

AIR PUMP *(see diagrams on next page.)*

This extracts the air, vapour and condensate from the condenser. The air present is saturated with low pressure vapour, and enters by leakage at different parts of the system. It is also liberated when the steam is condensed.

The pump is an inverted steam-driven unit, single acting and fitted with foot, bucket and head valves. The double acting steam cylinder gives the necessary reciprocating motion. On the up stroke, the bucket valves close, creating a vacuum below the bucket, allowing water to flow through the foot valves into the barrel of the pump from the condenser. At the same time, the water on the top of the bucket is discharged through the head valves to the Hotwell, the air and vapour being discharged to atmosphere.

This feed tank will also receive drains from various heating systems, and often contains filters which collect any oil or dirt in the water. On the down stroke, the head and foot valves close, the bucket valves are forced open, and the water below the bucket transfers to above the bucket.

AIR PUMP

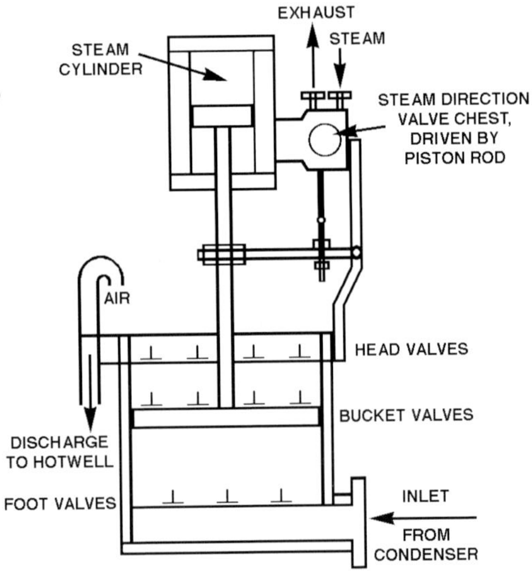

FEED PUMP

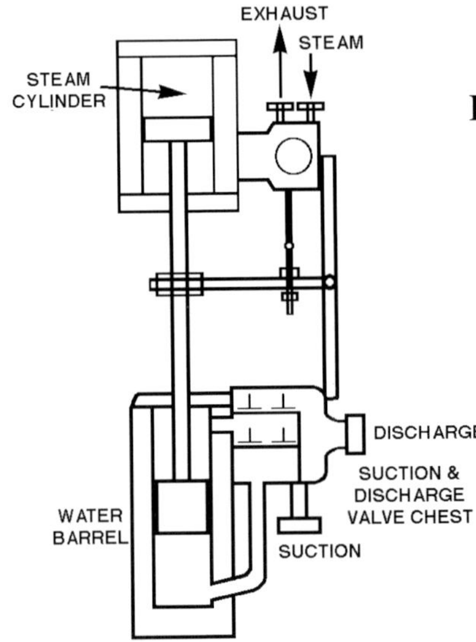

FUEL OIL BURNING

Preheated pressurised fuel is supplied to the burner tip which produces a cone of finely divided fuel particles. At an operating pressure of, say, 150 PSI and a temperature of 190°F, the fuel reaches the diaphragm where tangentially arranged ports cause a fall in pressure, resulting in the oil swirling at high velocity within the chamber.

At the outlet from the orifice, a hollow expanding cone of fuel droplets is produced. The film thickness is reduced with an increase in the fuel supply pressure. The burner output varies approximately as the square root of the pressure.

The burner is lit by inserting a lighted torch into the furnace. Starting from cold, no mechanical or electrical power is required. In many cases diesel oil is introduced by gravity from a tank well above boiler room level. As this oil does not require heat, a fire can be lit by a torch applied to a small burner.

Using boiler oil, the standard lighting up apparatus consists of a blow lamp heated coil through which a hand pump circulates the fuel on its way to the burner.

BASIC ARRANGEMENT OF FUEL OIL BURNING SYSTEM

(In practice, pumps, heater and filters are duplicated.)

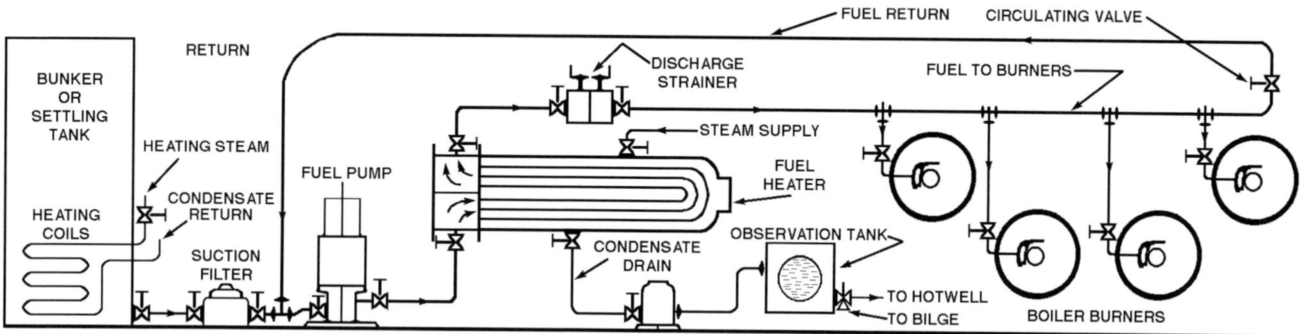

SETTLING/BUNKER TANK: Fuel is stored and heated if very cold to enable it to be pumped. Any water content is drained off.

SUCTION FILTER: Heavy impurities (grit, dirt) are caught in perforated metal baskets.

FUEL HEATER: The fuel is circulated within 'U' tubes, which are heated externally by steam. The temperature leaving the heater can be 20° to 30° lower than the flash point of the oil, e.g., with oil flash point 220°, good results are obtained by heating to 190°. Too high a temperature may cause the oil to carbonise.

CIRCULATING/FUEL RETURN: To circulate cold oil lying in the pipes through the heater several times to raise the temperature sufficient for ignition.

FUEL SUPPLY PUMP: Delivers oil to the burners via the heater, the pressure being at least 100-150 PSI The pressure governs the supply or output of the boiler.

DISCHARGE STRAINER: Collects fine grit in knife-edge disc which can be rotated by hand.

OBSERVATION TANK: Heating steam returns from tank and heaters are led to this tank which is fitted with a port hole. Should any oil leak because of a defective heating coil, it will be seen through this port hole. The water will then be drained to the bilge, preventing contamination of the boiler feed water.

RAISING STEAM FROM COLD: A separate diesel tank may be fitted to supply diesel, led into the fuel system at a convenient point, to a small burner.

STANDARD FUEL OIL BURNER

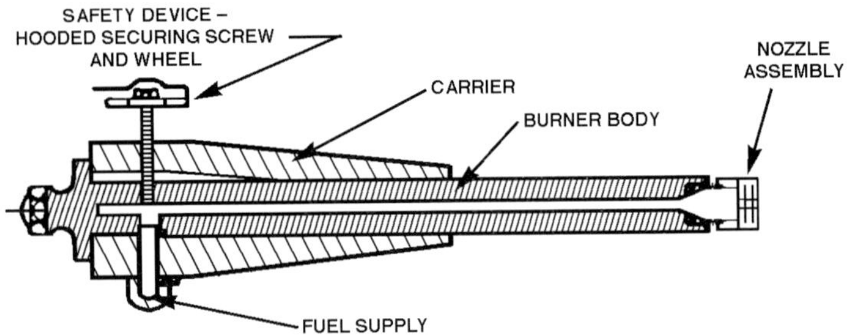

ASSEMBLY

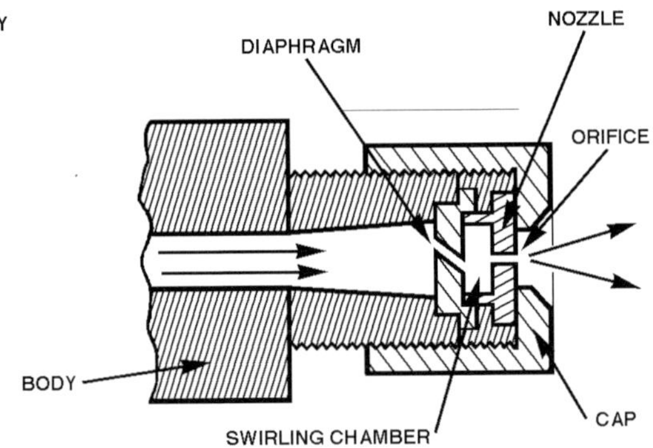

ENLARGED VIEW OF NOZZLE ASSEMBLY

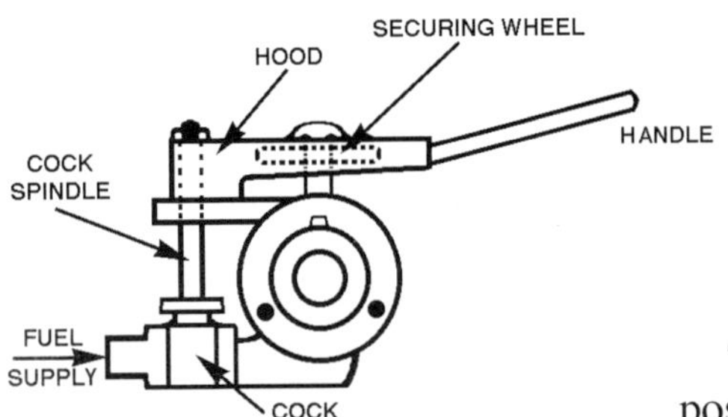

DETAIL OF SAFETY SHUT-OFF

The diagram shows the burner in operating position. To remove the burner, the handle is pulled towards the operator, i.e., revolved 90°. The cock shuts off fuel, and the hood uncovers the securing screw wheel. By slackening off the screw, the burner is released and withdrawn.

SPINNING CUP ATOMISER

This arrangement consists of a motor driven fan and a fuel cup. Fuel oil from the main system, heated and pressurised, enters the burner as shown, along with air for combustion. Due to centrifugal force and opposite swirl of air to fuel, the oil film leaving the cup is broken down to even-sized droplets.

Ignition is by electric arc from specially fitted electrodes.

Advantages:

1. Low supply pressure (15-40 PSI).
2. Improved atomisation.
3. Oil viscosity can be reduced.
4. High output possible.
5. Wide turn down ratio
   (This is the ratio of the difference between high and low fire burner output).

Disadvantages:

1. Number of moving parts and electrical equipment in way of furnace fronts.
2. Increased probability of failure, electrical or mechanical, compared to standard design.
3. Electrical power necessary to light up from cold.

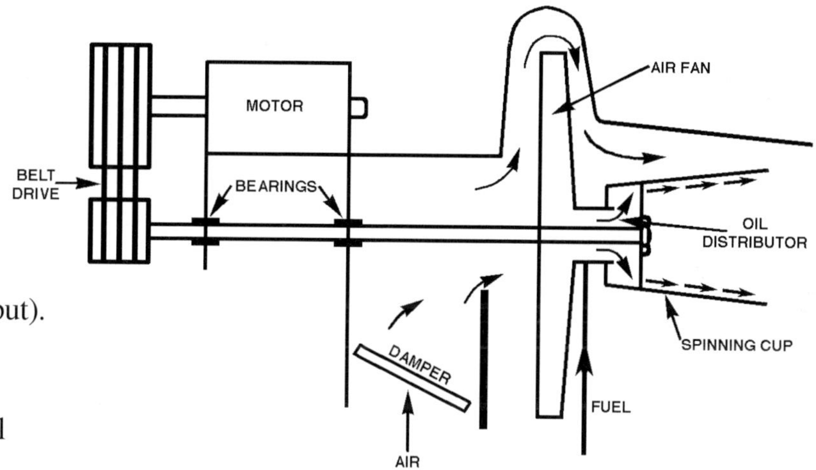

PRINCIPLE OF SPINNING CUP ATOMISER
(as fitted to ***Waverley***)

ENGINE INDICATOR

This instrument is used to obtain the Indicated Horse Power (I.H.P.) of each cylinder, at the same time giving conditions existing, such as the state of piston rings and valves.

The apparatus consists of a piston and cylinder which can be coupled to either end of each main engine cylinder via a switch cock and piping as shown in the diagram. The piston drives a parallel motion gear, terminating in a pencil. A spring whose strength is in proportion to the working pressure is fitted on the small piston rod. The drum shown consists of an internal return spring and a driving cord pulley. When the driving cord is operated back and forward, the drum will oscillate by virtue of the return spring. The driving cord operated from the reducing motion thus causes the drum to revolve in proportion to, and in time with, each stroke of the engine. Attached to the indicator drum by means of a clip is the paper on which the diagram is to be traced.

Thus the drum with paper attached revolves with every stroke of the engine, and the piston causes the pencil to rise and fall with the varying pressure in the cylinder at the corresponding part of the stroke. The combined movement causes the pencil to trace out the lines which form the "indicator diagram".

It will be seen that these indictor cards are extremely important if the machinery is to run efficiently, as they tell us of defects within the plant, whether valve settings are correct, and the horsepower developed in each cylinder. For smooth running, the power in each cylinder should be as equal as possible, and this is achieved by the expansion gear described on page 12 – adjustment of this gear alters the travel of each valve.

The switch cock, previously mentioned, has three settings, namely Cylinder Top, Cylinder Bottom and To Atmosphere. This last position gives the atmospheric line shown on each card.

PORTABLE INDICATOR DRIVE

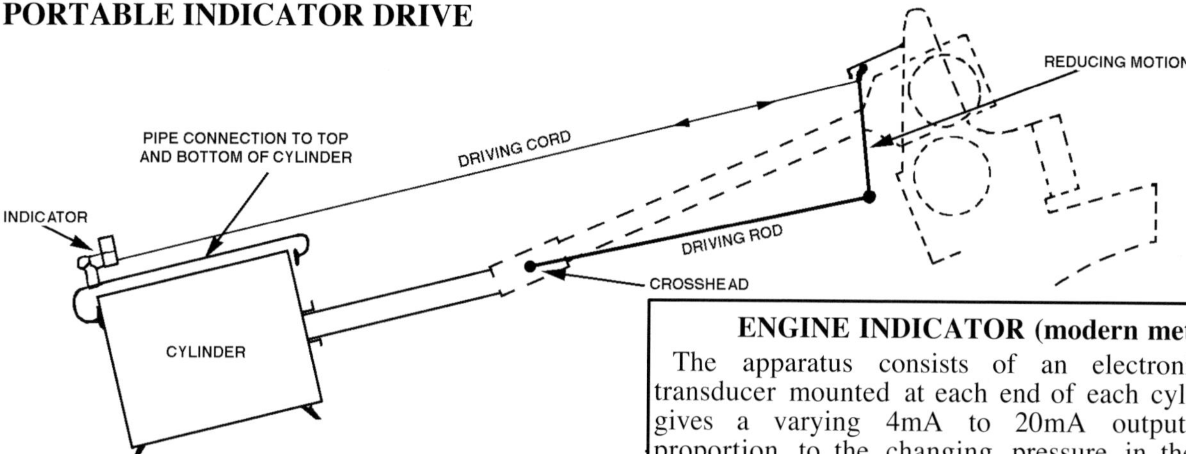

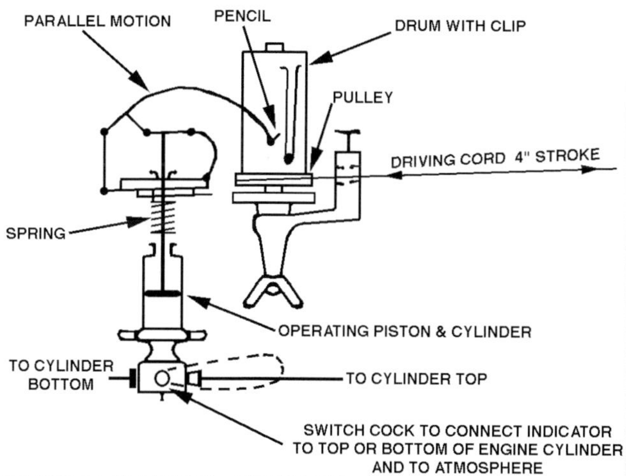

DIAGRAMMATIC SKETCH OF ENGINE INDICATOR

ENGINE INDICATOR (modern method)

The apparatus consists of an electronic pressure transducer mounted at each end of each cylinder. This gives a varying 4mA to 20mA output signal in proportion to the changing pressure in the cylinders. This signal in turn is fed to an analogue-to-Digital converter, the output of which is able to be measured by computer software.

The position of the pistons in the cylinders is obtained by the use of a rotary optical encoder positioned on the port paddle-shaft coupling. A reference pulse is generated every time the HP crank passes cover-end dead centre. All pulses from the encoder are interpreted from this point by the computer software, continually giving the exact position of each piston, all other dead centre positions and revolutions of the engine.

The software then takes all the generated signals and produces pressure-volume diagrams which are displayed on the monitor. These can be interpreted in the same manner as those produced using the traditional method.

For *Waverley*, at her normal cruising speed (45 rpm), the engines can be seen to produce approximately 400 I.H.P. per engine, giving a total output of 1200 I.H.P.

EXAMPLE OF CARD TRACED OUT FROM ONE SIDE OF H.P. CYLINDER

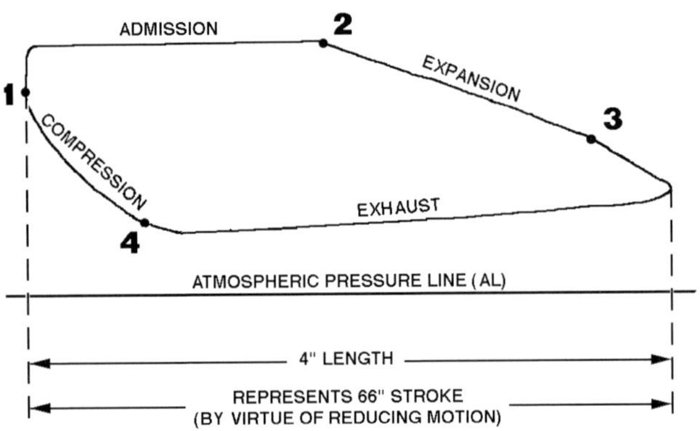

1. Steam admission, piston top centre
2. Steam cut-off, piston half stroke, going down
3. Exhaust opening
4. Exhaust closing

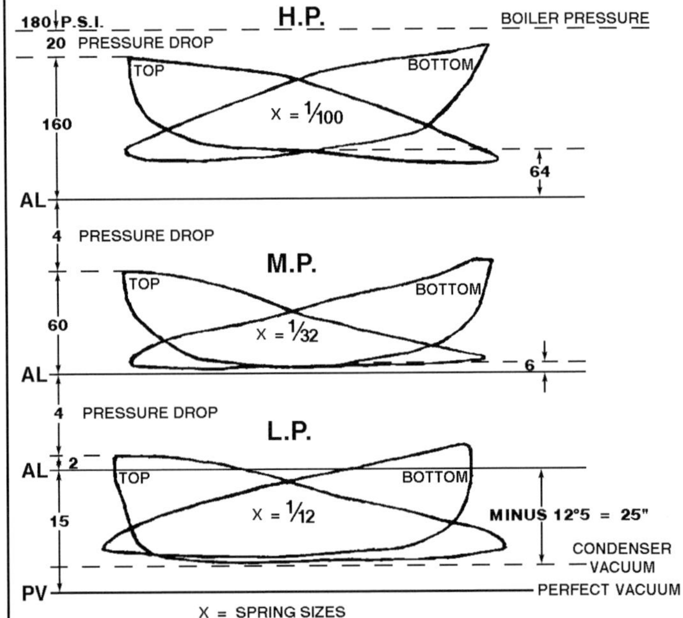

e.g., X = 1/100 means that every inch the pencil rises on the paper is equal to 100 pounds pressure, so 1 1/4" rise = 125 lbs, 1/2" rise = 50 lbs, etc.

DEFECTS

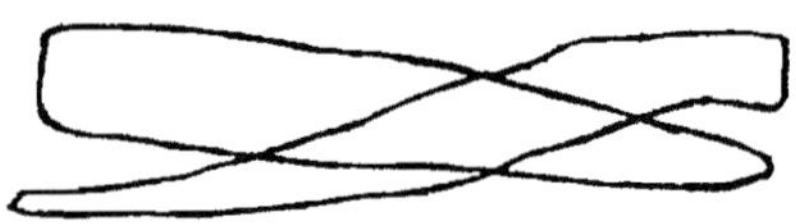

LEAKY PISTON RINGS

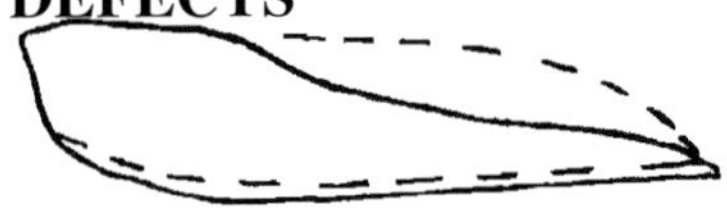

LEAKY VALVES

Steam is still being supplied after cut-off, as steam and exhaust lines are higher than usual (dotted lines.)

H.P. CARDS

SCALE = 1/100
SPRING

NOTE: Diagrams and figures are merely intended to illustrate the method used, so are not to scale, and do not apply to any particular ship.

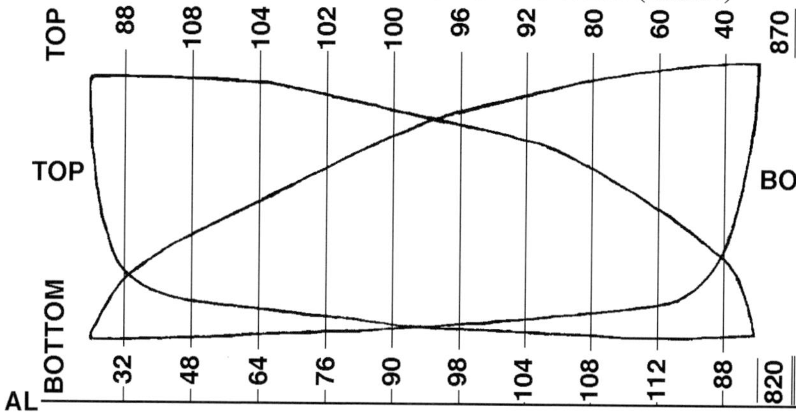

$\frac{870}{10} = 87{\cdot}0$

$\frac{820}{10} = 82{\cdot}0$

M.E.P. = 84·5

The **Mean Effective Pressure (M.E.P.)** is found by dividing top and bottom diagrams into ten ordinates. Measure each ordinate from steam to exhaust line using the scale corresponding to the indicator spring used. Record each measurement, and calculate as shown above.

I.H.P. Formula: $\frac{\mathbf{P \times A \times L \times N}}{\mathbf{33000}}$ where

P = M.E.P. (lbs/square inch)
A = Cylinder area (square inches)
L = Stroke (feet)
N = Number of working strokes per minute

For any cylinder, as only **P** and **N** vary, $\frac{\mathbf{A \times L}}{\mathbf{33000}}$ is constant. This constant would be 0·0754 for a H.P. Cylinder which is 24" diameter, and 66" stroke.

If r.p.m. = 50, then working strokes per minute = 100.

Therefore I.H.P. = M.E.P. x Strokes x Constant
= 84·5 x 100 x 0·0754
= **637·2**

In the same way, constants and I.H.P. would be calculated for M.P. and L.P. cylinders. The whole engine I.H.P. would then be the sum of the I.H.P.s for the three cylinders.

PADDLE WHEEL

The feathering pin is secured to a bracket bolted to the outer paddle box, its centre being placed forward of the shaft centre, and higher. The pin position is thus "eccentric" to that of the shaft.

The feathering strap revolves round the stationary pin. The "King" or "No. 1" rod, of greater strength than the others, is recessed into and firmly bolted to the strap, the other end being connected to the float in the same manner as the other rods. The driving rod gives the motion to the feathering strap, and therefore to the other rods.

The eccentric motion causes the floats to feather as the wheel turns, that is, their angle is altered to enable them to enter the water in a near vertical position, giving a good thrust. The float leaves the water in an inclined position, which prevents loss of power by lifting water up.

The wheel overhangs, and is supported by, a half shell bronze bearing mounted on the "Crow's Nest", a structure attached to the outside of the ship's hull.

Lubrication of the various pins is by water, with grease or tallow being used in the bearing.

MATERIALS

Wheel Centre, Eccentric Bracket and Strap	-	Cast Steel
Arms and Radius Rods	-	Wrought Iron or Mild Steel
Float Bearings	–	Bronze or bronze with nylon liner
Rims, Stays, Float Brackets, Levers	-	Mild Steel, 28/32 tons tensile strength
Floats	-	Steel or Wood

DIAGRAMMATIC ARRANGEMENT OF 8 FLOAT PADDLE WHEEL

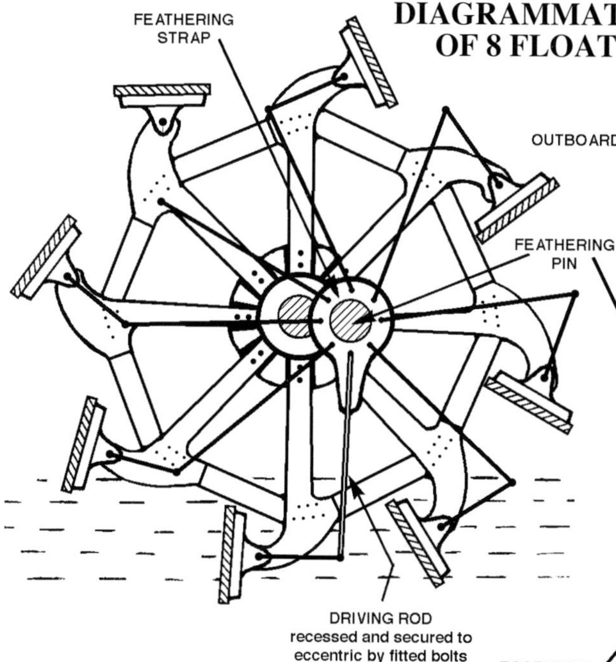

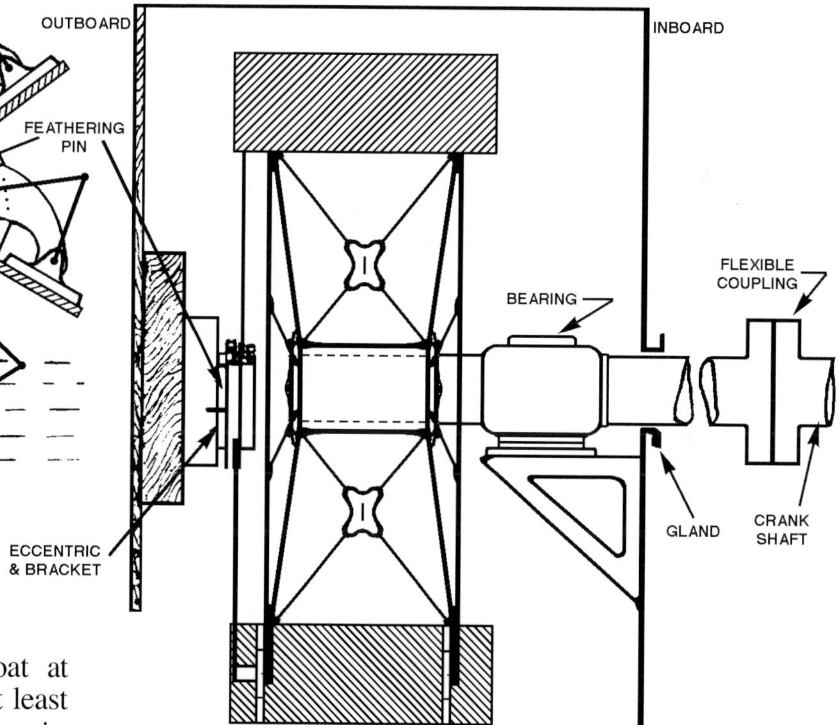

ENLARGED VIEW

FLOAT IMMERSION

The amount of water above each float at bottom dead centre is designed to be at least 1/8 of the height of the float. In fact in ***Waverley***'s case the immersion is 1/2 float height.

CRANKSHAFT FLEXIBLE COUPLING

(Showing one bolt in position)

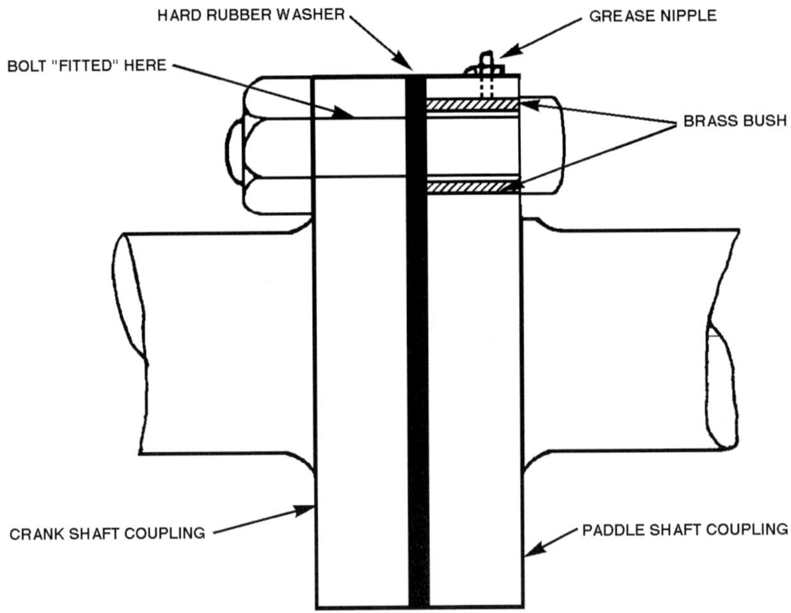

This type of coupling is necessary to allow for wear down of the outside bearing. The holes in the paddle coupling are made larger and a brass bush fitted to each, arranged to give a clearance between these and the bolts. This prevents the crankpin and webs being subjected to heavy stresses.

EMERGENCY HAND STEERING
DIAGRAMMATIC ARRANGEMENT

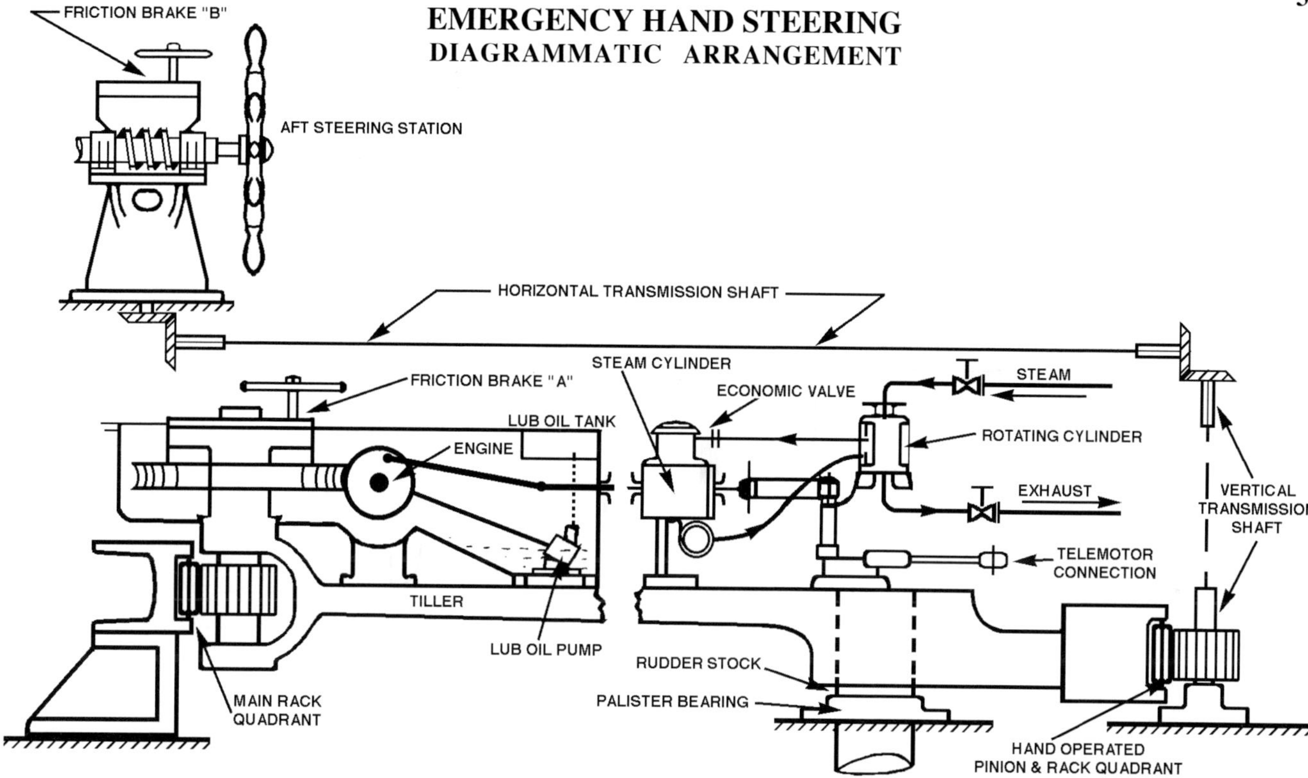

TO STEER BY HAND

1. Disconnect Telemotor
2. Disconnect Brake 'A'
3. Connect Brake 'B'

Steam or hand gear can be put in or taken out at any position of the rudder

ECONOMIC VALVE

Ensures a positive shut-off of steam when rudder is in the desired position. (This was removed in 1948 due to water build up in the steam line.)

STEAM & EXHAUST CONNECTION

A sub-divided cylinder which turns about the steam and exhaust pipes. The centre is in line with that of the rudder stock, whose action it follows. Steam and exhaust lines and valves are secured to ship's structure.

STEAM TILLER AND TELEMOTOR

The steering wheel, when turned, operates the rack by means of the pinion, so that the rams, when moving downward only, displace the fluid accordingly. This brings about a corresponding movement of the receiver cylinder. It will be seen, of course, that as one ram moves downward the other moves upward, receiving the displaced fluid from the other end of the receiver cylinder. The motion of the receiver cylinder is transmitted to the control valve of the steering engine, causing the engine to start in the required direction.

As the engine is geared to the quadrant rack and free to rotate about the rudder post, it travels along the rack and operates the rudder which is secured to the engine. At the same time the control valve is automatically readjusted to the mid-position (closed), thus locking the rudder at any angle of helm.

This arrangement is termed the "Follow Up" or "Hunting Gear" system, as the amount of rudder angle depends upon the amount of handwheel movement. If the wheel is let go, the spring compression of the receiver cylinder displaces the fluid back to the transmitter, which results in the wheel being turned back to mid position.

The fluid used in the telemotor system can be a non-freeze "Arctic" oil, or an anti-freeze mixture of water and glycerine. For efficient operation, the pipes and system must be fully charged, all air must be bled off and all joints must be tight. Working fluid is contained in a tank which is part of the transmitter pedestal.

For clarity, the emergency hand steering and details of steam and exhaust connections have not been included in this drawing, but are described on page 35.

BASIC DIAGRAM OF TELEMOTOR AND STEAM TILLER STEERING GEAR
(Similar to ***Waverley***)

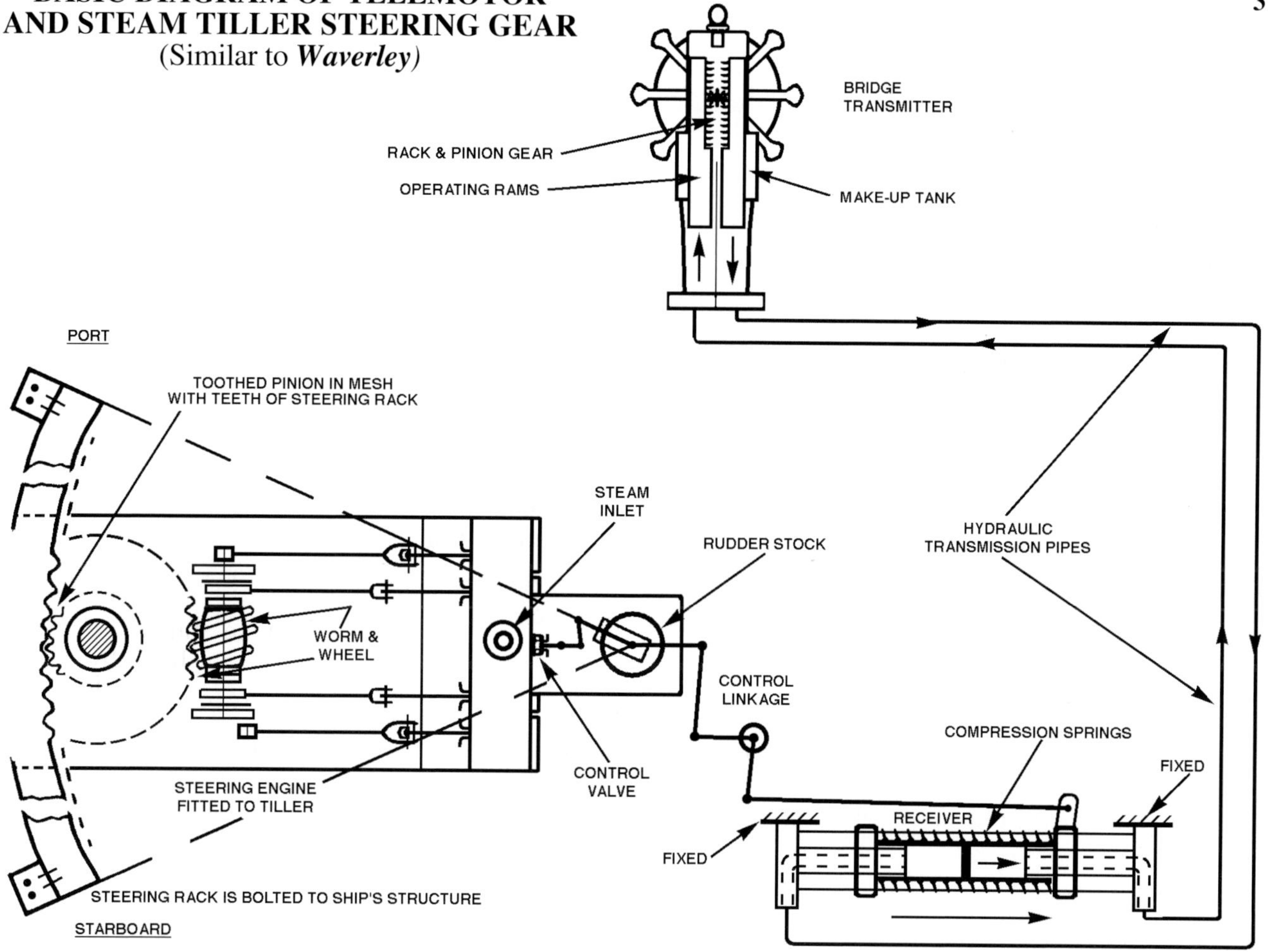

DETAILS OF PADDLE STEAMERS REFERRED TO IN THE TEXT

	Launched	*Broken up*	*Shipbuilders*	*Length*	*Gross Tonnage*
BRISTOL QUEEN	1946	1967	Charles Hill & Sons Bristol	258'	961
CALEDONIA	1934	1980	Wm. Denny & Bros., Dumbarton	224'	623
JEANIE DEANS	1931	1967	Fairfield S & E Co., Govan, Glasgow	250'	635 (originally) 839 (latterly)
JUNO	1937	1941*	Fairfield S & E Co., Govan, Glasgow	224'	642
JUPITER	1937	1960	Fairfield S & E Co., Govan, Glasgow	224'	642
KINGSWEAR CASTLE	1924	—	Philip & Son, Dartmouth	108'	94
LUCY ASHTON	1888	1951	T. B. Seath, Rutherglen	190'	271 224
MAID OF THE LOCH	1953	—	A. & J. Inglis, Pointhouse, Glasgow	191'	555
MERCURY	1934	1940*	Fairfield S & E Co., Govan, Glasgow	224'	621
WAVERLEY	1946	—	A. & J. Inglis, Pointhouse, Glasgow	240'	693

* Sunk on war service

DETAILS OF PADDLE STEAMERS REFERRED TO IN THE TEXT

	Engine Builders	*Cylinder diameters*	*Stroke*
BRISTOL QUEEN	Rankin & Blackmore, Greenock	27", 42", 66"	66"
CALEDONIA	Wm. Denny & Bros., Dumbarton	20", 31·5", 50"	60"
JEANIE DEANS	Fairfield S & E Co., Govan, Glasgow	26", 41·5", 66"	60"
JUNO	Fairfield S & E Co., Govan, Glasgow	25", 39", 61"	60"
JUPITER	Fairfield S & E Co., Govan, Glasgow	25", 39", 61"	60"
KINGSWEAR CASTLE	Cox & Co., Falmouth	12", 25"	24"
LUCY ASHTON	Hutson & Corbett, Glasgow (1902) A. & J. Inglis, Pointhouse, Glasgow	52" 28", 52"	60" 60"
MAID OF THE LOCH	Rankin & Blackmore, Greenock	24", 48"	51"
MERCURY	Fairfield S & E Co., Govan, Glasgow	22·5", 35", 56"	54"
WAVERLEY	Rankin & Blackmore, Greenock	24", 39", 62"	66"

THE AUTHOR

It all started when I was a pupil of Greenock Academy, then Dunoon Grammar School, where, amongst many things, I learned that James Watt did not invent steam or the kettle, but improved on the steam engine by adding the condenser.

In these days, many of us spent the month of July cruising on the 26 steamers serving the Clyde Coast, using the then popular "Runabout" Season Ticket. The cost was thirty shillings (£1·50) for one month in the late '30s, anywhere on the River on any steamer, except MacBrayne's and the Campbeltown & Glasgow Steam Packet Company.

Much of my time was spent viewing the engines, and sometimes I managed to get "below" on the ***Jeanie Deans***, ***Caledonia***, ***Juno*** (1937) and the ***Duchess of Hamilton***. This was the real start of my Marine Engineering education – I didn't know what I was letting myself in for.

Having got steam into my blood, I served my apprenticeship as a marine fitter and draughtsman with Rankin & Blackmore, the builders of ***Waverley***'s engine. Indeed, around late 1942 we received drawings of the machinery layout of the ***Jeanie Deans*** as a guide for a proposed steamer for the then L.N.E.R. Clyde services, once hostilities were ended. However, I went to sea shortly afterwards with the Royal Fleet Auxiliary and sailed on the Russian convoys and the Western Ocean. Later on, during leave, I visited Rankin & Blackmore and saw the engines of ***Waverley*** and ***Bristol Queen*** under construction in the shop.

All my working life has been spent deep sea, except for a spell with the Caledonian Steam Packet Company, just when motor vessels were being introduced to their fleet. During my years as Chief Engineer on passenger and cargo vessels, bulk carriers and tankers, I have seen unbelievable changes, not always for the better.

Steam has given way to motor, automation, and the most tragic and indeed dangerous of all, the decline of our Merchant Navy. Through this, redundancy caught up with me. Not being ready for full retirement, I offered my services to the Waverley organisation. Being accepted, I assisted in the refitting of ***Balmoral***, then after a spell aboard ***Waverley*** prior to the 1986 season, sailed as Relieving Chief. I am now happy to continue with this arrangement, including working-by during the winter months.

I have tried to make the drawings and descriptions as simple as possible without being too technical. My thanks to Leslie Brown, Gordon Wilson, Derek Murray and others who have compiled the data given into book form, which I trust will be of assistance to all those who are really interested.

R. J. Ramsay, C.Eng., F.I.Mar.E.,

Chief Engineer

Greenock 1988